//:universal attraction upon inference of talent that makes unbalance seduction.

Inferecing that, like the othere, one can resist the flame that strecthes a million light year.

Unifying both the stardust and the spectacle. Switching faces and turning them back to the origin of Cosmic future. The letter love is not empirical because you switch the meaning.

It now is name that which wasn't there before.

The amor fati is love of Zarathustra

He admits that love is selfish and wants it by himself. He admits that love wants to be more predictable and stable. But it also admits it wants a future where no-one is more like him.

If love was eternal and not Zarathustra then life wouldn't make sense.

Every book ever written by Friedrich Nietzsche was a composition of what it feels to catch himself never letting his sight fall.

Even after that he finds himself. He knows that life is a game of struggle against tyrants.

Nietzsche has one option to defeat tyrants in the same composition.

Nietzsche has to become more devious and with more armor against enemies of revenge.

That is why I Spoked thou to you to surrender your pity.

Your grace is upon the entrance where the hidden holds a wishing well.

There you will find a reason to nuture the well and send thousands of wrolds into the Universe.

Nietzsche said it was impossible but he made it possible.

The unfinished work that provoked Nietzsche to fling into the abyss and return as soon as the abyss got tired of Nietzsche.

Nietzsche upon a dying star he flung like a mad man and struck the clock into the mid point where time and space separated into a new clock where time would repeat the event where the Universe started like a turtle who forgot if time was made from sand or from dust.

But the ending that started the Universe was increasing for Superman to create his time machine.

" I cannot destroy the Universe because I would be destroying the ultimate version of myself."

Every star in the galaxy is a reminder that sunlight is not the only matter in the universe.

Even space dust is composed of life.

The very beginning of life there was no life like earth because life was not made off skin-cells.

Life was made off tissue that resemble the crocodiles.

Every species in the habit of land is composed of knowledge including the animals.

I happen to see that the Universe is getting larger than the Ubermench.

I expect to Surpassed the Universe as soon as possible.

Every landscape is in the mist of loosing a possible ancestor.

I come from the celestial tribe called Tritan.

The average man conforms to reading and writing the opposite prefers solitude to withther his thoughts bringing a sense of reason, his most desire is to destroy reason. For he can't bare a sense of relieve, even for Zarathustra that was his non saying to life.

How should I proceed today in my solitude.

In away I could create a mirror that reflects my inept desire to create.

For creating is wondering a place where the less fortune and the more brave are more incapable of defeating.

And yet I have not met a person that could not fly because his time to fly was not perfect.

If a man wish that he could fly he would first need to learn how to walk then learn how to see.

Every person is like a child he sees before he runs.

That makes the child run slower.

In order to run faster, you must run faster than what you are able to see.

If you lose you win a new sight.

For Friedrich Nietzsche he said that you can win a race only if you believe your race is superior.

I did not know a race is won by power and not by weakness.

If you want a strong race you need your abilities to determine how you want to win the race.

Your determination in conquering is compared in how you let your reason feel and tolerate.

If you want to conquer a girl who is the opposite of your

desires.

Then wish a better woman who is more like a landmark, someone who can see her deepest desires.

The secret to finding this landmark is upon the hardest task.

A woman desires to be free and to be with no secrets.

In her eyes the perfect man is a myth who made her believe that no men is ever a woman.

Until Superman who was more perfect than the deepest pharaohs and who tremble at the sight.

Superman is beyond the landscape that governs the sea world.

Up until now the sea monsters have come in place to bring forth a wive.

Someone who is different than Superman because they fear his strength.

For every person who was born incepious has learnt that he has no desire to be young.

He thinks that old people are scared of dying even if they are scared of going to heaven.

Not even Nietszche was that inspid to believe he was immortal.

Friedrich knew he belong with the strongest people of all history including Napoleon who was a Monster.

For every person who was born incepious has learnt that he has no desire to be young.

He thinks that old people are scared of dying even if they are scared of going to heaven.

Not even Nietszche was that inspid to believe he was immortal.

Friedrich knew he belong with the strongest people of all history including Napoleon who was a Monster.

Under the impression that you hold, there are more or less ways to seal into the ocean. The impression upon your distance, how we try to get far as possible. We come back in the ocean that left us in tremble and seduce the very wave that brought us back.

I fear the tide wave ocean where I won't conquer the land and the night sky.

For every night the ocean falls back into the night it came from.

Even for me I speak how I love to burry the death in their treasures.

But mostly their treasure dies inside their corpse before ending.

I once forgot how my life took a sea monster as joke.

In the jungle where most fear to sleep I slept there with the moon facing my back while I slept in front of the night.

//:everywhere the intent to take power into a field as if the power was lacking, and yet the sense of power makes a disguise for someone to take it.

I believe every form of ignorance drives power away making power take itself.

That is why power deserves a new form of power.

Every idea is composed of will the sound of a monster.

In any situation comes from mistake, the mistake of power going into a cycle where reason is at fault with nature.

In any sense Friedrick taught nature to be senless to power, to order power to be powerless so that nature could live in a cycle.

There are no laws of nature that prove this cycle except the return of Nietzsche who comes back as a cycle.

This proves that nature wants to be in power with the spirt, and yet I want to be in power with the nature of Nietzsche.

As soon as I become Ubermench I become part of nature, this allows my sense of reason to be eternal in every possible way including the cycle that has no intentions of ruling.

I believe I can nuture the cycle back to power, where nature would be incharge of life.

That is if nature doesn't overturn into a monster, that is because I have not yet nuture the monster I want to become.

And only because I lack the skill, I do not know if any monster should lack nature in any form of reason.

I have yet to know how much a monster is if there was a person like me, and yet I do not know how to progress this idea in any possible way.

It seems in any way I can surpass the ability of reason as a tool of danger, but I cannot manage how reason is above my rank of power. I put forth the idea that I am invincible but I have yet to tested out in praxis. If any I wish to know why my abilities decrease as soon as I meet danger in a unfamiliar scene, as if they take my power out off my conscience and take it to my conscious state, there is no way putting a mental stop at the moment, but I believe there is a way to put my power into the hands that took it out.
If only I could remember the days of my training, the way I used to when I felt introuble against my will to power.
I need to be invincible for the sake of my sanity and stability, in order to defeat the entire strategy.

It seems like Nietszche was not very clear on how to take the truth out of context and put it in the same manner, the only truth that people want to hear is there voice, it makes them sound incredible, including people who worship themselves as a tool of revenge and yet can't even figure out if they are not making a mistake by not being evil.

They assume they have what it takes to be in power over someone like Nietszche, and can't put forth their despair and weakness aside the fact that Nietzsche is made out material they wouldn't understand, I believe every individual deserves to be in a sense powerful for their sake mind, but when you try to confront a Monster for being a Monster.

People are going to question their values as a whole, even denying their conscience of what is true.

Every individual is a the peak of remorse, their life's have no future in knowledge, and yet they wish they could become more.

The very last attempt to be different than Nietzsche was to betray his deepest desires and overturn them into a mistake, as if I was the one controlling their values. I only put forth the only escape they had of reality, I even taught them to be more inhuman and yet they thought they were playing my game cards against me like if I was not fit to be Nietszche.

How I despised the ones to be my rulers as if I wanted a ruler.

I don't know why I have to be commanded to be less than a master.

I only wish to be in control not doing the opposite of what is commanded, and yet my little voices that tell me who I should be, makes me sound like a lunatic without parol, and I'm not insane but I am totally out of my mind, and that is because I think I'm not yet respected as myself. Everyone can play to be Nietzsche but in the final stage they turn againts each other like if Nietzsche was the one who turnt againts the Germans.

I believe I didn't turn on myself or any ally because I would be a disgrace to be Nietszche.

Even the most highly trained spy can not out run the danger of power, the very most important aspect of Nietszche is to be Ubermench in every single possible way. Yet there are people who have said that the only mistake I ever made was snitching, I didn't do the favor because I was caught in between the conflict.

I overturned the conflict and they took me as a threat to their affairs in war.

I do not wish to be In a conflict that I can not win, therefore war was not an option but a mistake, I put forth that war is not inevitable but it is possible to despair the power to integrate the power back to the source.

I only wish power was not a artifact but a tool to control the situation againts tyrants and rulers.

As a monster my purpose is to make power not take it away, I am Superman not a tryant who is weak for power.

My only purpose is to revenge the way I was brought back.

Looking into the mirror in sort of a strange way I wonder how I became so common looking like a dark pale color and yet the feeling of another appearance feels almost the same, as if I've transformed into the thing I was trying to avoid. I yet to see how that happened within a life time but I know that the origin is nowhere in my appearance, I have travel across the land looking how evolution wants to be, at best it is a disguise to infiltrate, to look at people and be almost identical.

Even for Nietzsche we was identical to a German Philosopher who reminded him how different he was.

I have seen that looking into the mirror I am not a human as I appear but rather the opposite.

If what I am to be a Ubermensch then also I feel that in the years pass I have gained the wisdom to know who I am in the most disturbing way.

The color of my skin Is a reflection of who I was when I was a kid, the memory of what I thought was normal was only a instinct because I was not in my original appearance, it took decades of evolution to disguise as a local as if I was the only kid who didn't look good enough. I now understand that color is how you socialize in public, but in the higher ranks that's how you get power.

I used to be in the Era of Nietzsche the higher rank, I was the highest Philosopher the world has ever known.

How I became the opposite of Nietzsche is what made me stronger than Nietzsche. In Norway they call you foreigner but in Dutch they call you Aryan because it is common to see how a person can not be Aryan if they are not from Europe, and yet Europe is not only a country of foreigners but also a country of culture, the culture that brought light to other countries in the world.

I identify as a race that never took the appearance of the Aryan, but I come to conclude that race is not what's the Aryan, it is the Aryan that makes the race.

I will not stop believing that in the future the race I was to become will be inherited as a gift.

I can only gain the advantage in looking different.

I have taken into consideration on how people behave towards me, as if I was not worth being near them. They point out the worst things they can think of and then they feel as if they have expose my true identity, as if they want to be in control of something, and yet I have no pity for such accusations for they are not in any way worse than what it means.

I think that in the future they will not remember their hate but remember the feeling it gave them. They want to be remember as the good people who stood up againts a monster who was more cautious than appearance.

I want to know why they feel as if they can put up with such an act and feel like they invented the thearther show.

As if they think they can in return pay me back with the worst gestures.

I assume I am not the only one, people also mistreat those around me to empower their system of revenge, and yet they have not taken me serious because they are not capable of doing such an act.

The only way to stop their stupidity going further, I will need to assist myself in treating them worst.

The intervention of their master plan is out of style and is not getting better, I for once am told they feel humiliated for being less intelligent than myself.

I told them they were only humans and can't compete againts a monster if that what you want to call me.

I make myself clear that I gain no power in their defeat, I have yet to explain why I will never take them as my opponents because they will never reach my perfection in evolution. They will not gain wisdom as much as the wish, I will not be humiliated to their despair ignorance.

They will not be powering the world with that set of mind and they will not be in control for too long if they keep pretendending to know their objective.

I have made clear I am not their enemy but it seems they want to face one, I will not try to be their equal for that is adsurb and disturbing, yet I have not seen people tell me who I want them to be instead of a monster.

I have notice how people see me as a enemy, they think I took their freedom or that I took their liberty. Yet I have not told them why that is not possible for me to make such arrangements, the freedom was taken by someone who was not me or in anyway my choice.

I told them I was not even in that situation. I was in the wrong place but not in the wrong side, if any liberty was never given to you by your founding fathers, they didn't believe people had equal rights and even so I believe equal rights are not equal. Furthermore the more you point out that I'm a traitor to your nation for speaking on behalve of your liberty. I was turning myself less human in become more human, I despised my inhuman form but I gave me power and now you despised my power.

I wonder if you chose your leaders or the the leaders chose you to be their servents.

I have no reason to power a nation who has no business in returning favors.

I know you feel betray by some maniac who didn't care for rights or laws, but Nietszche said he was a diplomatic and he felt rights were earn not given, Nietzsche assume he was so smart he said the right to bear arms is a right to murder any person who is wish to be corrected.

I cannot correct Nietzsche, he was my master in the process of creating myself, but I know those rights you go fight in war and feel libertarated is just another way of saying how ignorant you are to speak, the right that came to you by default was not original or in any way true.

I feel like you need more rights to protect your status but the status is already under attack by the government who wants you to be more ignorant about your choices and your decisions.

I proclaim myself to be a different breed to gain control over the silly mistakes you think of me.

I wonder if Friedrich ever wanted a companion, someone like himself with the same destiny and experience, the person who could have made Friedrich feel wanted, and yet I have been unsuccessful in finding what I believe to be love. In any case I struggle to comprehend why so much has been missing, I have concluded that girls are interested in men who are well behave and have a sense of humor, I therefore lack those qualities in praxis, but nevertheless I have tried to reason with myself and have discovered how girls love to be taken out of their comfort zone. They whispered how to love them but that seems like a great task, even after the burden of proof.

I later say to myself if is it worth the trouble and the danger of looking for a companion.

Eventually girls love how they love reason and love the ecstasy of finding a perfect partner.

I have yet to look in the eyes of those that know how it feels as if nature was looking for a way out towards.

The common question people ask about is if I die?, what happens in return?, I assume the world needs to be careful if they want to know how the world gets informed about the future. In any case the world will not be staying in contact with any form of intelligence outside of earth, this is the case with Wagner when he showed how music transformed the vocalist in the Era of Nietzsche.

Eventually I get how the invention of music changed the way we think of sound, every note played by Beethoven was instrumental in knowledge, causing sound to rationalize the vocals of the Era, and pherphaps fewer mistakes made by pianist. I assume you know more about the structure of the universe than the structure of knowledge and reason. Furthermore my music should play out into the vast where sound doesn't escape the galaxy and with sound of a monster it should stay hidden for eternity. That is why my music is the heard people want, they want to heard in the scope of knowledge that plays music in the tone of Friedrich Nietzsche, I for once am the only singer with no vocals instead I have instruments that play in tune with universe. Instead of asking how you play with no vocals, ask how you play with no music, I ask if any music is music to my taste and take the sound of your garments and vocalize the structure of the universe.

In connection with how much I stay hidden among the most horrific stories is a mystery that even Friedrich Nietzsche couldn't answer for himself, even after the war of Europe he fought the regime of tyrants and fought with himself, denying his true strenght, which made him into a monster he could not control. Every single moment was less beautiful and more insane, causing him to visualize the devil as the work of Nietzsche, who made the devil a monster that spoke in the voice of Nietzsche, that is why Nietzsche spoke to me in his true nature. He spoke to me " In nature you have discovered the secret, you have given me power over thee and you must embrace your new power, I grant thou my armor and shield, may you know more about power my dear companion."

Go fight in the dessert to be more evil and more dominant towards race, you need to be made of sand and look after it as your mother, the time has come to be more than water, take your time to reach solitude if that's what you desire, never reach upon the eye of horus or you will lose your eye, if any case don't smile back to the dead or you will die too. Never learn to forget the worst, in case of memory lost never loose your eyesight, always take your thoughts with you in case you loose in battle.

Never remember the lost or the found.

Take good care in finding the lost but never take care of what isn't found.

Eventually the most harsh adventure is the way people see reality, they think it's formal and consistent with thoughts.

I deny reality to be anything more than a mirage, because you desire the most things you wish you had, the illusion of sensation, I assume I sense the very thing I wish to be when I'm not Zarathustra and with the notion of transformation I summon the Superman the most impeccable force of importance up until the death of Friedrich Nietzsche, I was summon to be a Ubermensch, like the son Nietzsche once had.

The most Intelligent adversary that was ever made, I took myself for a monster and now I am the reflection of Nietzsche who was more insidious and terrible than any human being.

It Is upon the hour to seek a solution based on inference, commonly known as resolution in as much as too much resolution is made of the same material as reason, the old saying that fights monsters in nightmares that can't be experienced through resolutions, I fear Nietzsche won't comeback in any nightmare I resolve, because God knows he will be forgotten by Christians.

In as much as I support the intention of what looks to be almost important to the senses I prefer the less important, the only way to get stronger in any case of knowledge it would need harsh conditions of which have never been attempted to those mortals, I summon the empty dialogue played by Prometheus, his imprisonment was in opposite of Nietzsche who fought outside of reality, even so much was lost that in the spectacle of loosing, everything broke including Prometheus who was never a slave, partially a god that knew the most disturbing truth.

Even if I was to become strongest of any living being, I would have to break free of any partial truth.

I thus seek out my wisdom to thou take me into the most forbidden truth.

I almost forgot how to speak like Aristotle, his insipid tone where he took logic into the voice of Nietzsche who was more careful in context, giving out the premises before the idea was even taken serious, after the great philosopher of Greece, Aristotle was the most sophisticated logician in ancient Greece, I therefore owe the knowledge that was put forth in fighting the tryants of modern day politics, as if Nietzsche needed Aristotle to that himself, but I got stronger because of Aristotle who gave me the ability to master Nietzsche in a single day. I therefore know more about the secret hideouts that Nietzsche never spoke of and that Zarathustra was not just Superman but also a visitor of another timeline where he worn Nietszche the future about politics and world domination, I for once I'm told Zarathustra has given many gifts, my gift is to see the future events of my Era even after I long pass away in the most epic battle, the truth may be told that Zarathustra is no stranger to my world and he has comeback to see my most epic story of all time, I thank myself for inventing a time-machine where Zarathustra comes back to see me in my most powerful state, even after the battle I will be waiting for Zarathustra to return with more gifts and maybe in this case, the gift to see the past before future.

How much I miss Nietszche telling me how I would be the Superman, in the long run I forgot how I made myself the monster I am, how deceitful, how cruel was I turning towards myself, I felt no pity for myself, I was not human and in return I failed to past my limits, for what seem impossible to note, I was in the very top, no one could have reach a certain limit, not even myself for I was not alive or dead, I was a living tissue, looking inside for meaning and found deep consolations, what gave me this imspicable apparatus in which I never saw another form of logic and thou saw me naked like a tear drop falling inside a concave in which I was so impressed, nevertheless I was not in impress with the idea that love was endless and that Nietszche died lonely because he was depressed, instead he was very brave in love and never gave up hope that some day he would find his most beloved wife, for he was very optimistic in love affairs, his romance was so much like a child who learns how to breath and his favorite song was the sound of rain that pounder the sky like the way mountains feel after snow falls. This very moment I realize that Nietzsche was more insane than I could ever imagine, his look, his stance was more like a sea monster looking for air in the ocean where there was no hope of survival, I get how I look more or less like Nietzsche but how I miss the hope of looking for a sea monster that could take my hope of survival.

I now see that I am very vicious with hope and that makes me less alive inside reason, take my knowledge for what it is and not for who I want to become, for the latter I am already a Superman but for the former I am not a intoxicated depressant in need of narcotics, I for once I'm not a narcissist but I wish to be one for the sake of adventure and desperation, in any case I don't like to feel numb or feel impotent.

I only use was necessary to resolve any solution that needs to be done out of pity and disgrace.

If any feels better is to stimulate the sense of power that I have taken upon myself as a Ubermencsh, I believe to be enough in the long run, if that is the will to power.

I have taken in consideration how I have made no mistake in being a titan where there was no escape in the bottom of the oceans and taken the lives of those withing the limits of survival, I have no intention to be a sea monster but I do wish to be a monster in land where there is no escape except the ocean where it is blue in color and white in salt, but nevertheless I have no wish to reopen the case of Nietzsche who fought againts the evil of those who could not speak like him, he thought of many ways to reform the speech of his conscience and lost many battles in trying to remake his works, I failed to see that for myself.

Friedrich was not a creationist or a evolutionist, his idea where life came from was informal to many spectators, his wish to be irrelevant to the case was seemless, he mentioned life was a mistake of God or that God was a mistake of man, in any case the idea that I know where life came from is not important to know because life is not a living matter, but a living tissue made of silk and not just silk but organic silk that we see in space, my skin is also silk but not from space, it was given to me by the goddes and queen Cleopatra as a gift of power, in return I gave her my soul to take, that was our deal in the universe, to steal the richest form of live and give it to the powerful for enterprise and fortune, if I may speak In behalf of Cleopatra she was not so like Nietszche, her humor was not absent and her taste for revenge was impressive, I almost forgot how she dressed, but I think she wore a silk like vest and a silk like dress to match her incredible eyesight, her eglea like stare and evil like smile she was so impressive with her people who thought she was a goddess.

In as much as I try to figure out why a technical problem falls in some disorder, I feel impelled to fix a solution that isn't worth solving because technically a solution is more difficult than what it actually is, in different perspectives the solution is solvable becausee it was never a solution to fix the problem.

I made no error in trying to know how to solve a technical problem, in my perspectitive I didn't solve solutions that weren't possible to know, I therefore know two things about solving a technical problem, one is that solving a technical problem cost a certain amount of effort including knowledge of that effort, second every technical problem is not solve by mastering the solution it is fix by mastering the problem.

In any case I seem to mistake, I make no effort in knowing how to solve something that isn't possible to know by effort, in any case I failed to see the effortless mistakes that one has not yet experience in any field of science, that is why carefully looking back I made no intent in trying to be more knowledgeable, I however was never interested in knowing why something work the way it did, that is because I was not in a perfect motivation and therefore lack substance in figuring out the motive, if any how I was not in good conditions to conduct a repair that was not given to me by instruction of integration, that is why I lack effort in mechanical motives, for I have not yet seen the pattern of reduction, and is therefore not lacking in my philosophy of intelligence that I have put forth in my explanation of how the universal structure works, in my defense I am not a mechanical engineer but I am a mechanical infrastructure which is different than most mechanics that I have ever known of, and that is why I feel more relatable to any system of structure known to space.

In the process of inventing a solution I made no intent to know why I bother trying if the problem was never the issue, the problem was the solution for my sake of sanity and it took me a minute to know why I sounded so desperate in my relation to something meaningful, I have not yet seen why I failed to be more like Friedrich Nietzsche who never questioned any sort of relations attach to emotion and despair, I have to be the first imbecil to know why I need a invention made out of philosophy or what purpose would life have if any of my solutions were already solve, in what way was I perfect than not imperfect, I made no intent to know the imperfection of my doing, and yet I have no answer as to why I need to be more perfect than before, like if I were not in any sort of way a new person, I believe I have made no contribution to society and thus know why I am not good enough to be reliable to any sort of will that has not been given to my power, I will to power the very most delicate things that I have yet to experience, and yet I have no idea why I have the power to know everything that has structure and intention, but who am I to question the most intelligent beings, I have no clue who is my creator if any one should exist or if I'm just a person living in the most horrific scene ever experienced to mankind.

I need to know who is my inner most trusted alliance so that one day I can forge a outcome to my enemies and friends, I have learnt to know more than I speak of but that is why I have never spoken of what wasn't possible to know.

"I bring home to you a gift said Zarathustra", "the gift of symphonies, please accept the gift into consideration." I very much will appreciated such gift to you my dear friend for now I have no issue in providing information of what symphonies are in musical alignments, this instruments have been playing since the beginning of time. A symphony is a note played in many notes to symbolize structure in making more symphonies therefore the one that represents the most sound in a single symphony is the most recognizable in all the melodies, the most disturbing sound is the fifth and the seventh symphony, because it makes less sound at the end of the each tone being played, therefore if you look carefully you can play a singe melody in just two symphonies and that takes away the instrumental value of each symphony that has more or less value towards the audience.

However attempts I make I make no effort in making mistakes, I love making the best effort possible for future generations to know that it takes more than knowledge to know something, it takes courage to make more, even after the odds of finding more is almost impossible to comprehend, I have lecture myself today in hopes of reason, I have left many people wondering if any possible structure of power is possible, I have no clue if that is possible or if not impossible, I have given thee the hope of power to make and to take, I have no reason to know why I know so much, I'm just that good in any situation and feel almost honor to be taken seriously, for I have yet been dishonest with myself in trying to be more than Nietzsche but I am honest to be like Nietzsche, if any I am not greater or lesser than Nietzsche, but I am somehow superior to Nietzsche, for I have shown many what I have accomplished with so little time and so much to talk about, even Nietzsche couldn't make my music sound so good, the music you play to people who hear the meaning to be different, I have not yet reach my limits and hope to catch more feelings along the way, if any more should be sayed is that love is a matter of intention and intelligence not a superficial intention on how to be romantic, I hope more to come if any come back to my senses for more than a song.

Horrific I think, what is horrific to Nietzsche, his music I fear, but my music is more horrific, because you can't compared music to horrible music, I made the distinction on what it means to be horrific with music that was never the case with Richard Wagner and Friedrich. The testimony is in between both constructions of relations not bot both constructions of relatives, I have spoken both my truth and my friend Nietszche, but I have never spoken of a truth that was not truth in syntax, if anyhow I was promoting Nietzsche in his final words of power who I have power the monster I wish I could have been, but in latter I was more monster than Nietzsche for what I have done to myself and to my friends who are not helping stabilize the situation at front, that is more than enough to communicate a syndrome of remorse and better yet to believe dying is better than living a life of fear and disappointment, I have yet not been very clear with what I have not done to surrender those tears people wish I had, but better yet I am the most powerful infrastructure known to mankind.

Every now and then I speak to the devil because his afraid to be lonely, or I talk to him to feel better at night and it sounds crazy to feel better than the devil, but he knows who we are and how we play our cards to mankind, if the devil speaks he wants to know more about me, he cares to know if I have the courage to carryout orders that given to me, I say yes I have the courage to carryout any order I desire to be, but I also know you want the best warrior there is, I know he commands like Nietzsche but he also disobeys my request to speak about the future, I have no reason to know the future but I wish I was more evil than you, Friedrich is so powerful he gave me full command over his army, I have no intention of hurting the weak as if they were useless to me in the first place, Nietzsche had more to tell, but you my dear companion you must carryout the orders as commands or you will perish like Nietzsche who was fighting the monster he was.

I assure you my dear companion that we are invincible in transformation, better yet I wish you the best of luck in transformation for that is what we are a beast at heart and a monster at best.

I want to know why I can't transform into the Aryan beast if that is what I am?

Maybe I need more time to relax the temperature of my body, or maybe I need more intuition on how to behave, I have yet not seen results for over four years and now I have waited too long to see my hair turn blonde like Friedrich Nietzsche, I assume he made me look like a freak of nature to see my full potential or maybe I was not trying my best, I was lacking intimacy with myself, or I was not in contact with the solar eclipse, maybe I was born too soon or maybe I have lost my will to be Aryan, in any case I am now prepare to transform back into that beast, if any one thought how that was possible, then hear well because music doesn't play twice.

The color blonde comes from the ancients who transformed because they were not powerful enough to fight wars, they were taught to dye their hair blonde to look superior but failed to be permanent, that is why Zeus the son of Cronos, gave blonde people the ability to feel beautiful in color, they were disgraceful with the gods, so Zeus order blonde people to have darker hair, in the mist of finding true love.

That is why hair color is different among people of culture and that this appearance is only permanently because one day your blonde the other day you are not.

Don't take yourself too serious with dying hair because it only serves one purpose and that is to feel powerful.

The temporal effect Is in effect, the deficit that in return comes by reason, not involuntary or by will, instead it is in effect the turning point between good and evil, I proclaim both to be the same effect, if only if as if only the truth was that simple to memorize, or that the truth is will to power, but what Nietzsche was in effect saying was how you will not be in power by pure will, or how willful the transparency is towards the involvement of knowledge, if as if only there was no question of how disturbing the effect looks alike, in simple terms effects have no intention to rationalize the true spirit of reason, that is why the temporal effect is causing dementia in different people who have forgotten the effect of time.

Friedrich was right, he was not so tough againts my will to power, he showed more toughness through writings that were simple to taste the melancholy of the modern man, even so the so called higher man who ever in fact more or less stupid to know anything that wasn't more distasteful than sounding more like priest, who made no mistakes in their doings, but were conservative when dwelling in church, I sound like Nietzsche on the fact that he was right about everything including the sound of the old bird catchers, the symphony of melancholy, the terror that spoke to those in need of revenge the Christian apostle and the Christian Saint, who were in connection with nothing except revenge againts the god of Zeus and in revenge with the Greeks who were not monolithic, the sad truth of the most revengeful, yet the most hateful, how to be in revenge with sense of living, the idea that death is better than life, how adsurb is this opinion, like if God wanted to kill every single being for being sinful againts their desire to be more powerful and noble. The most disturbing ideology that has yet been put to rest by Nietszche, who I have summarize in return of that vengeance. I will not fall into despair for living the most iconic story ever told in history.

I have return more insane then before you freaks of nature, who wish to put my soul to rest in a place where God fears to entered, I have broken my chains of despair and I am now free to write the most involuntary manace ever to be put down in earth, I have shooketh the rock, I have stared space into sleep paralysis and now I remember that old ghost that came to me at night for revenge over my soul, how stupid I was feeling for feeding my soul to her slumber, I feel ashamed of who that really was, it was my mother being in a custom playing to be the witch of the 7th floor. I have no idea what her purpose was in reality, just that she tried scaring my soul to God who was waiting for me on my knees, I feel almost ashamed to admit how she got me fooled for a second.

If only she was more like Nietszche she could have made me into an Atheist much sooner than God, I wish to recall my despair moments where I was in no need of God and that somehow I was so relieve to be away from church, now that I think of it was more astonishing at how they pull it off.

First they pray before sleep, then they play a song in the middle of the night making it look like it was just another sign that a ghost was coming to sleep on my bed, then they acted like I was fully asleep to put a heavy object on top so that I couldn't scape the nightmare, I was sure it wasn't a nightmare, the question is how does a object so heavy fall on top of your shoulders? The answer is simple they put you in a sleep comma where you feel like you can't breath or scape the terror of dying, I was in some form of gas paralysis in which I was injected by some medical apparatus, the most disturbing sign I ever experience was the sound of death that woke me up in the middle of the night.

In relation to emotion I have no clue what I suppose to attract in connection to my feelings, I have never really question how I was ever different to myself, I was hoping that one day I could include my knowledge inside my reality, that perhaps I could be emotive about my relations a bit careful and take pride in doing so, I soon forget my distance and look for new emotions, if I'm not careful I will fail my mission, the impossible mission where emotions turn right upside down, what exclatly am I trying to be emotive? Perhaps my disinterest in logic and my boredom that hunts my thirst for knowledge, I may perhaps need to venture more into the abyss and find meaning inside, I wish I could turn my emotions at will to empower them in their most valuable moment, the extacy of feeling more inside the spirit as I feel like a Ubermensch, the question begins if I want to be emotionless or be disturb by emotion in so many fractions of reason.

If anyhow I find emotion to be irreplaceable I will not be willing anything to my desire, I would undo my will and take a different approach to reality, if that's the case I summon Friedrich Nietzsche to disrupt my attention in the hopes of alternating my reality into more questions.

I have therefore learnt to control the way I feel in the sense of reason, which helps in canceling the alternate reality I have summon in place of Nietszche, that is why I have no intention in finding myself among men or saints, because I have not question myself if emotion is a addiction to power or if power is an addiction to emotion.

Friedrich Nietzsche said that woman were created in consolation to being a mother, I have therefore question who is the father of the baby put forth in this notion of knowledge, if any how I know how many answers I can answer is a miracle I have not discovered in my sleep, I have therefore not answer myself if the father is truly a father of heaven of if the father is truly a father of hell, if I know that I am right in the moment than I know that I was in fact certain of what I am to be more than a god, but less of an angel myself, I have not mention if any possible truth is possible to know in reason or if I have taken this out of context, what I seek to know if the baby of a mother a baby that was created into a womb by a male reproduction system or if by chance the womb was created before any conceivable genital contraption, or if anyhow I have lost the trust of my judgement and I'm making an inference on why genitals have less ability to create the born baby into a unborn baby, if by logic that sounds impossible I assure the opposite is true and that genitals do comport to some degree a true born baby based on mother organs, I conclude that born babies are not made of genitals or womb fetus, but is made of womb genitals that contract into a fetus at birth of contraption, therefore I forget who is more alike? If any male is corresponding into genitals then by chance babies are not both female or male, the genitals have a degree of reproduction that generates both systems of organs and what makes a female make a baby is her fetus reproducing himself inside the mother while the father looks both ways to see if the baby is both mother and father apperance, I have now answered where the baby is born in the most disturbing manner I could think of, but I will not tolerate impunity over gender indifference upon relic information and transvaulation, I guess the most important thing to remember is the chance of birth that is given in seconds of living inside a cell, I mentioned that in praxis I have no intention to resolve a solution that isn't made clear in my paragraph but it's important to note both the mother and father have the less chance of giving birth to any child alike, and if this makes sense to science then it makes no sense to philosophy, for I have not mentioned the motive behind the reproduction, based on what motive there maybe in the unborn baby is based on knowledge that the baby wants to be born in a certain aspect and is therefore expected to procreate into something similar to the mother and father, I have not seen this in action only in theory that I have thought of while being in bed, but I have not seen results into why father and mother need babies to survive in the world of insanity, I have no reason to question evolution or any scientific fact that wasn't present in the moment but assure the fact that babies are not made to survive by themselves and therefore need more nuture than any living creature on earth.

I was hoping that by the end of the day I could write like a seafarer looking for ways to dive in, I am not looking close enough to see the mountain in the fright of sound, I am close enough to see some monster attack, but not to worry I won't get bitten by any creature in the mist of my despair, I am almost certain I can fight in my sleep like if I was some sort of space monster, I am probably out of my league to be a monster with wings but I can fly without any wings, I have proven this for a fact in such so many forms, I am sure this is more like a remake on how the end of hours made Cleopatra feel, as if they were not in love in the first place but that in my scape of drive or die, I was more insane then I could think of, I was shown how to be me in the most epic imagination of adventure, I will prove this right here in my writing that flying out the window of any floor is more save then flying out into your dead wish, I am not certain what it is about cars that drive by themselves but I drove my car into a ditch and lost control, I think I can be more cautious next time, in the future where there is no one stopping me and drive like a taxi driver, faster than need for speed or any game simulation at the moment, I am certain I was not in my true form of intelligence, I was partially awake in transforming what I ought to be in the future, I am not the driver I am the only conductor of the driver, I will not be pass as fool, I will be the one who knows the rarest scape possible for the benefit of action and drama, that gives me hope to look after my interest and best intentions, I soon will be the very best driver that I can be in reality and take over the wheel that puts the drive in reverse.

I want to know the dinifition of noon-tide? Where does it go? Where does the moon go? Into the tide? I guess I will not be certain to know more or less about the issue, I have promoted a theory of what it looks alike, in theory I assume that it means the moon going down the ocean and never escaping the bottomless waters, I guess I'm not scared to know when noon-tide comes to mind in the bottomless sea world, I have conquered the most destructive theory but never like this one, noon-tide the end of night and beginning of darkness a place where people disappear and take a different fate, I almost lost my life to the night fall, but I insisted otherwise I looked into my soul and found no life left inside, I can say for certain I have escape the deadly waters of the cold Breeze, the forgotten land of Atlantis and the unforgiven promise made to poseidon, at last thus found my shoulder in my head, I found the last Alantis the under sea monster, he forgot to be named after me but in return I came from a world of night falls and soon after I was relieve to be back in land with the people who opposed the night fall.

Taking inside what needs to be sayed in terms of what doesn't sound right to people, the worst idea that has to be made is made possible by composition of letters and thus invoke the most ideal landscape that could be nuture, I question my knowledge of the necessary requirements to surpass my interest of what I believe to be impossible to question at best, I wonder if any ideal invention is worth noticing in context of the most ideal, I am sure I can surpass both strategies in a single hand battle, I know no limit to my ignorance and pronounce my self interest as a necessary lesson, I am not imperfect in knowledge, my knowledge is imperfect to my reason, I cannot contribute more ideals, for what ideal is less wanted of myself, I become more inventive by reason than by knowledge, I want to be more upset than my worth is value, I wish to know how to be less evil towards my peers and know that I am more insane to begin rationalizing defeat in order to know what is to be taken care of after myself, I am at best the worst inventor of ideals, but I am not worth the trouble.

Insane upon time, what I'm a to expect both, what is not truth is almost both, thou speak to me languague of nature, I am not hearing you speak, I am not hearing you roar the sound that frightens the gods, why because playing by nature is chaos to those in power, I have not made it clear to know why I speak like a monster of eight light years away. I have to transport my thoughts into my myth, my dear Nietzsche you were not playing hard enough to be heard right, I am just going to speak a little more about us, together like we used to think in nature.

Please hear the sound I speak of at the sound of treatment, I only wish to be In power to know what the old saying used to say, I have incarnated my voice, more loud enough to fear no one in my sleep, I have yet seen my voice get a little darker, and thanks to myself I will only sound out loud what needs to be heard, I apologize if it sounds like a ferious lion gone mad hungry, for invention of the heart, but I know it feels the need to be heard more often than not, I fear I may scared the voice that goes tick tocking in my hour of sleep, I sound like an insane monster gone mad wild for the need of help, I am left with no alternatinative as to who is my enemy or who is my friend. I'm not hurting myself speak louder, I helping myself speak stronger, please use my idea to help you sleep at night.

Breath my soul, I have no need to rest, I have no wish to be less, I am not your ideal or ideology because I have no intention of roaring too loud to scared away sleep, I just want roar a little to scared of the most disturbing sound nature has ever pronounced in hitherto, I am just no playing word games or translating any single word I can think of, as a matter of fact I'm sick of hearing nonsense to my beautiful ears, I have no business in hearing what I need to know or what I need to learn from pressure upon reason, I am not the kid you expect to like as a friend of need, I'm not a kid that needs to know what a kid is useful for, or what is not sympathical.

Night approachest the deepest remorse, for one night I feel more alive than before, I have taken the liberty to get up with the sun and take the fall with the rain I wish I was, now remember who is not superior to myself, take a pill if you are afraid to know more than the less you desire to feel, have no pity to know more than you are capable of considering, have no intention to loose any battle that isn't for the win of power.

I feel no pain in trying to forget my instinct nature, the will to power is in my blood, I have not issue the best theory yet, I have yet to experience any will to kill in self defense I, but I will not hesitate my dearest wish to know more about intelligence, I have no reason to be more than Nietszche because we are more alike in dealing with power structures,

I tell you that in my prime moment I have not yet seen any will to power besides the fact that I am not inferior to knowledge, I used to play serious when I was a kid, but now I have forgotten how that feels, I feel no intention to feel any will that isn't my own, I will not disprove my innocence towards that which is my dearest hope.

Hope who is not a artifact that can be forgotten through depression and intention, I am not a inbemcil to any hope I wish to be In love with, I just assume hope is less attractive than real knowledge and dreams of forgetting my mission towards the ideal, I have yet been missing the real feeling of intention towards my ideas, please understand that I'm not insecure about nothing more than reality. I based my apology towards gratitude and not weakness.

Deceitful is the one that plays more with my thoughts than those who play the drums, I have no idea what people feel inside their bodies, I guess they know more about the autonomy of the weak and powerless, I know the true intention of the strong, they want to be in line for something more bigger than their ego and have no idea what a ego actually is for, if I may explain what ego is in relation to Nietszche who was more knowledgeable about the human body and the wisdom it carry inside the soul, I know that the ego serves no purpose in surviving any attack made by oppressors and despite the warning they think the ego is wise to be weak in self defense, I know woman feel the need to be weak in favor of the strong, they can't concentrate, they are blind in hope of forgiving the strong who have no idea what power is to them, they feel temor, knowing the feeling of regret, I feel no regret in feeling afraid of nothing.

Incredible intuition I feel when I get in more love, if that's how you remember, or if I remember who knew how to kill a mocking bird by the stone that I lay upon the ground, anyone was more invested in power than myself, I felt in touch we nature, I wanted more intimacy with a lonely bird who was lost in the woods, any feeling of hope was enough to trap myself in a cage of nightmares.

I repeat nothing more adsurb to myself interest, I guess loosing enough power is simple, but getting stronger is much more simple than it looks, I have no reason to feel afraid of nothing, yet I look straight at nothing as if I was not In terror to know more, I have to push forth my intelligence to the next level, I have not given enough thought of who is listening my inner child and thus speak more like Friedrich Nietszche who was more insane than lucifer, but who is in my most disturbing awaking in the sorrow of pity, and my disguise as a bird catcher, I am not in that mood anymore, I am more or less feeling weary of my song that keeps telling me how much I need to improve my physical appearance, I should say I need more time to structurize my memory a little further than Nietszche, I am not in anyway more stubborn than any missing piece of advice left behind the ashes, I need to be the very best for that lonesome reason, I can't comprehend any more logic at this point of departure, and soon enough I have lost my voice of insanity that has now become normal for my peers to listen in silence, I know they feel like hope is keeping the spirit up in awake, I know I can't sign the song I love the most, I miss the old Nietszche who was not bored to speak up in different tones, I know what I must do to keep evolving more energetic in my stomach and gut, I have reason to keep playing music that sounds mechanical or instrumental, I have yet seen my mistake in dealing with power that was over the edge of my sword, I know power is most wanted in moments of despair and hope, If any feeling is inside my head is the tone that keeps making my soul dance in motion with knowledge.

What exactly drives the will to involuntary action in self defense, they think they have a stronger echo in their ears or perhaps they feel the same tone of music that drives them more inside towards a window that scapes sunlight into their life's I assume, or they wish to be echoing something that isn't normal to hear, even for bats the echo is much more simple than repetition, of some idea that comes to them in some form of instrument, I believe this is consplaying my sound of reason, they wish to be more in tune with what is heard, in return they pay no attention to something obvious but cry out wolf in the streets of Massachausetts, I believe to be no more than a cry for help, they can't accept defeat in the most honorable manner, I guess they feel evil enough to be simple enough to laugh at the face of what is dangerous in nature, I assume they have no idea what it feels to be more powerful than any living mortal, or as I say in my tone of voice, they are not in motion with my conscience and can't unwind the most terrifying sound spoken to my ears, I believe it hurts to know more than usual and is not a thing to play with, I believe there is no more hope given the fact that they loose power in hope, they gave me no reason to be less than I hope to be, I am not saying it is easy to be a mistake of nature, but that nature is not a mistake.

What is a Superman? I asked Nietzsche in my remorse, he said Superman is not what it appears to be in any situation, he is not made of any living matter known to mankind, he is not even human or man in that category, he is not even a alien from another planet, he is not evil or good, he is both, I know Nietzsche is crazy to say he knows more about Superman than the actual Superman living today in modern day, so what is the secret I've been trying to hear all my life? Superman is my heroe said the most beautiful woman, but who is my future wife? I know Nietszche has a problem with people who want to be white because they feel superior enough to earn the genetic material to be noble in nature, so who is my wife supposed to be Friedrich Nietzsche? Am I alone in the universe by myself to experience nothing but power, I am not hoping to know why I need to be alone. I need more intelligence to know why I have to be Superman, please answer Nietszche if you know why?

Any monster capable of hearing at sound made of wind, is more monster than monster, I am single made by knowledge, not by power do I know everything, but by will of knowledge do I power.

Friedrich Nietzsche I have not yet seen why I know more intense music than fellow peers, but I have yet seen your music fly into my soul of terror, excuse my innocent languague for my tone is just getting used to hearing the voice of my conscience, I fear I may not know how to spell the way you taught me, but I know how to respell my song that put up with, in case you know why I sing the way I sing, it may be due to my inferior motives of revenge that I get when I get angry at myself and my surroundings, yet I have never seen anything like any other song ever played into my soul, I expect to keep learning the musical high note that I speak when I'm inside my deepest thoughts, I have pronounce a keeper that is no yet seen in any form of learning, yet I know I speak for myself when take matters that are serious in nature and power, yet you have not seen the last word of the measurements that I've been thinking when I read my language of Zarathustra the goal keeper that sold a record inside my remote control, I pronounce thus my word to be superior to any cycle of life that has lost it's meaning in the impotency of your inferium languague, I pronounce that I am at least close enough to hear my voice sound deep inside.

Zarathustra in voice of my will, I see that I have yet mastered the gift of reason and knowledge, but I do know that I have seen more intense equipment inside my deepest thoughts of knowledge, I fear I have not seen your face inside my voice, but please use my knowledge to hear more things that are not impossible to describe in nature, I know more can be said alone in the darkest hour of need, but I used to know more than what I used to say, I have knowledge of the most reasonable intruder, that has yet seen light into my soul, if any soul is possible in this universe, or if the universe is a soul maker to make no sound of knowledge so that I drink the dew of my despair, I have never seen a lost soul feel deep inside the trenches of apparatus, and thus know more equivalents inside my remote adversaries who have forgotten the logic of Zarathustra.

Friedrich Nietszche why has my enemy spoken the truth about the truth, if any truth is true enough, why should I know more about the truth that has no sound, or voice, yet I spoke like if truth was more evil in natural effects, did I know something about the future past, or was I in a day dream talking about love affairs and love notch, what exactly is love upon reason, I love therefore I know? Or do I love therefore I know that I love, what is true nature that was played to me in the voice of freedom, I have no freedom choicing my destiny if destiny is freedom of another destiny, what is the sound that I must hear in order to know? What keeps playing my song, the song Nietszche wrote for me, I have yet seen no more, but I have seen more or less, why was the music so infatuated with my voice, I have no way of knowing what keeps my mind up at night, except the sound that played me out as if I was looking at my own demise, I have taken my self serious enough to know I want to be more in connection with my own knowledge, why was Nietzsche playing a horrible song that resemble my most inner wishes, I taken this game far too far to be less real than what ever dream off, I know if I dream further I will break more than one illusion but the whole truth apart, the inner most desire was kept a secret up until Nietszche said what needed to be said, and thus I have no idea what want any more than what actually need to do, I have played the most disturbing sound in the track of tracks and now I don't know what is not a real context, I know more about myself than I know about Nietszche and have concluded nothing in respect to knowledge, I have yet told so many truths about Nietszche I forgot that I share those same truths, I must keep silence about myself, I must not hesitate in going out in public domain, and I must respect my wish to be discreet about what actually needs to be done out of knowledge, if I hesitate to know more I will lose what I've started, I know I must keep fighting my innerself, if I must exchange a truth for another truth, then so be it.

I must let go what needs to be done out of power, and proceed to the next step, I am not in control of destiny because I have seen the most disturbing truth about myself, I've seen what most wanted to see, the monster that was covering a beautiful surface, and spoke of things that impossible to know, I care to know why I was discovered, did the music know more than I could think, or did I just loose my inner child, for a better understanding of what I must see to know how to feel better.

Tell me more, what I need to hear, why makes me need something that isn't my creation, better yet need to know more, what is a need if not a want, but I need to know why, yet I need to know more than I need, because it is not possible to will something that isn't part of my need, I need to will I don't need to know why I will something out of power, yet I need more than power, I need to know why needs to heard, I know I need something more than power, I need myself not just my will to power, I need to to know why I need to know more than I need.

I have no need, I have no need because I know why needing something isn't worth my time, I know my time needs more power than I need to live, but why I need to live if life is not worth living upside down, upside down is not a will to power it is a need, I need to be upside down looking for my eternal revenge, that is the need of hour, the turning upside down to feel like I need more, the hour of need is not a need it's a must have to know more, I need to be on time or else time ends my need, I need to made of steel to know that water is made of the same material as liquid is made of rock, I need more water than steel because water is more expensive than building, yet I have not build anything out of water, I need more steel than water to be more intense, I need more water than steel to be less intense, I need to be made of pool like water to be less intense, I need rock material to be more intense, that is why I need.

Friedrich Nietszche why thus I feel like a hungry hippo, if I know why I am not a hippo, what makes people think I need more air inside my belly, I feel like I need more ingredients not so much intestine, but yet you know if people ask that I feel hungry why must I negate such truth about nature, if a hungry hippo is not hungry enough why must a hippo be a hippo? I must answer my own question alone in the dark shadow of thought, I feel no need to feel hungry If I am not angry enough, I must eat careful.

Nietzsche has more immense testosterone but fails to used it completely inside a fermux, I know you keep making a song more complex than that of doctors, doctors are not equals or resequals they are more tense, to see if a theory works on patience than of themselves, I wish to be taken care of by Nietszche if the case was critical to make power inside my sickle cell membrane, born to live thus born to die, but not hope, that hope stays inside, thus I hope, for the worst I have not yet seen my eyes spark hope into my chest of treasures and I fear I can't see pity inside walls of knowledge that spell out nothing outside.

Nietzsche how wrong were you, you were my friend and now you are my enemy, you betray my knowledge. How ferious was your intent to forget? Just so you could relearned again, I have no clue if I want that mistake, but thanks for the clever mistake, I thought I was learning more from you than from myself.

Thou Nietszche you know why mistakes happened? Because intention is missing, you know that feeling you get when you are more intense, well that is awful, because it doesn't help the power grow, it weakens and then fails, it fails again, and you were not meant to fail, but were not scared to feel any more, that is why you must have succeeded in trying to win, because you are more powerful than any person I have ever encounter in any land of the seven worlds.

Remember how knowledge play that beautiful song? Where you must not mistake it, well that mistake is over, now we learn, we learn to know, we learn to learn if we want to be more like you or if I want to know more than myself, don't get it wrong I am not myself when I'm hungry for knowledge.

The more instrumental that is played, the more you know how make a song play itself, thus you are more instrumental, Nietszche used both instruments, the voice and the note, but failed to sign the full song, that I played in my backyard as a kid, the kid that knew how music was connected to something not normal to any mortal, yet I was in full control, or so I thought, the mortals know music that plays that is it, but they don't know how instruments are made to help sign a better song, thus spoke Nietzsche who failed to be more intune with knowledge and reason, he thought reason was impossible to reason with, but not impossible to understand.

I forgot if knowing is not leaning properly what I need, or what I don't need, I forgot what I'm supposed to be doing, now I have not yet seen nothing more than my ignorance towards my inner desire, my door has left my window, I feel more realistic with my inner most impontence, yet I fear nothing, yet they wish.

I heard what I needed to hear, thus I know why I why, I keeping playing music that no is synonymous but it keeps playing around, it feels alive inside my gut, I feel more intune to know more.

www.ingramcontent.com/pod-product-compliance
Lightning Source LLC
Chambersburg PA
CBHW062305290526
45794CB00006B/2701